St Croix Falls Library
230 S Washington St
St Croix Falls WI 54024

CLIMATE CHANGE

CLIMATE CHANGE AND POLITICS

BY MARTHA LONDON

CONTENT CONSULTANT
Barry Rabe, PhD
Professor of Public Policy and Environmental Policy
Gerald R. Ford School of Public Policy
University of Michigan

Cover image: Some people hold protests to try to persuade politicians to address climate change.

Core Library
An Imprint of Abdo Publishing
abdobooks.com

abdobooks.com

Published by Abdo Publishing, a division of ABDO, PO Box 398166, Minneapolis, Minnesota 55439. Copyright © 2021 by Abdo Consulting Group, Inc. International copyrights reserved in all countries. No part of this book may be reproduced in any form without written permission from the publisher. Core Library™ is a trademark and logo of Abdo Publishing.

Printed in the United States of America, North Mankato, Minnesota
082020
012021

THIS BOOK CONTAINS RECYCLED MATERIALS

Cover Photos: Shutterstock Images, foreground, background
Interior Photos: Noah Labinaz/Shutterstock Images, 4–5, 43; iStockphoto, 7, 12–13; Red Line Editorial, 10, 22–23, 35; Clara Margais/picture alliance/Getty Images, 15; Fabrice Coffrini/AFP/Getty Images, 16; Siam Pukkato/Shutterstock Images, 19; Richard B. Levine/Newscom, 24, 45; Andrea Comas/AP Images, 26; Paul Morigi/Getty Images Entertainment/Getty Images, 29; Aude Guerrucci/Bloomberg/Getty Images, 32–33; Ronald Patrick/Getty Images News/Getty Images, 36; Charday Penn/iStockphoto, 39

Editor: Marie Pearson
Series Designer: Katharine Hale

Library of Congress Control Number: 2019954184

Publisher's Cataloging-in-Publication Data

Names: London, Martha, author
Title: Climate change and politics / by Martha London
Description: Minneapolis, Minnesota : Abdo Publishing, 2021 | Series: Climate change | Includes online resources and index.
Identifiers: ISBN 9781532192722 (lib. bdg.) | ISBN 9781644944257 (pbk.) | ISBN 9781098210625 (ebook)
Subjects: LCSH: Environmental management--Government policy--Juvenile literature. | Climatic changes--Law and legislation--Juvenile literature. | State and environment--Juvenile literature. | Environmental policy--United States--History--Juvenile literature.
Classification: DDC 363.738--dc23

CONTENTS

CHAPTER ONE
A Climate Emergency **4**

CHAPTER TWO
Recognizing a Problem **12**

CHAPTER THREE
Piece by Piece **22**

CHAPTER FOUR
An Uncertain Future **32**

Fast Facts 42

Stop and Think 44

Glossary 46

Online Resources 47

Learn More 47

Index 48

About the Author 48

CHAPTER ONE

A CLIMATE EMERGENCY

Outside Hamburg, Germany, people filled the streets on September 20, 2019. They marched toward the city center. A stage with microphones was set up. Many of the people held signs. Some chanted. The people wanted world leaders to create laws to fight climate change.

The people in Germany were not the only ones taking to the streets. It was the Global Climate Strike. Millions of people around the world marched. People on every continent

People in New York were among the many around the world who participated in the 2019 Global Climate Strike.

YOUNG LEADERS

According to a 2018 study, 56 percent of people 55 and older were concerned about global warming. In contrast, 70 percent of 18- to 34-year-olds were concerned about global warming. Young people also were more likely to support fighting global warming. In 2019 young climate activists pressured Democratic leaders for a presidential primary debate focused on climate change.

joined in. Young people led many of the marches. Swedish climate activist Greta Thunberg participated in the New York City march. At only 16 years old, she had already spoken to many world leaders about climate change. The Global Climate Strike was one of the largest global protests ever. The message to leaders was clear. It was time to take climate change seriously.

WHAT IS CLIMATE CHANGE?

Climate change describes a difference in Earth's average conditions over a long time. The climate of Earth changes naturally over time. However, human actions have caused it to warm at a faster rate. This rise

Burning fossil fuels creates energy that can be used to heat homes and provide electricity. However, this releases greenhouse gases that contribute to climate change.

in temperature is called global warming. It is due to greenhouse gas emissions. Greenhouse gases include methane, carbon dioxide, and water vapor. The gases reflect heat that would otherwise escape back to Earth's surface.

Human-caused warming began with the Industrial Revolution. During this period from the late 1700s to the 1840s, people created many new technologies. Transportation and factories led to an increase in the use of fossil fuels. Fossil fuels include coal, natural gas, and oil. People burn these fuels to create electricity and to power vehicles. When burned, fossil fuels release carbon dioxide. The gases enter the atmosphere. They build up over time. They trap more heat and reflect it back to Earth. Earth's temperature warms.

EFFECTS OF WARMING

Climate change has serious effects. It makes the global temperature rise. Glaciers and ice caps melt. The ocean warms. Rising temperatures stress plant and animal life. Warm ocean water also causes stronger storms. Climate change harms people too. It leads to air pollution. Rising sea levels flood islands and shorelines. People are forced to leave their homes.

Scientists warn that people need to keep Earth from warming more than 3.6 degrees Fahrenheit (2°C) above temperatures recorded in the 1850s. Beyond that, catastrophic climate risks become far more likely and difficult to address. By 2019 Earth had warmed 1.8 degrees Fahrenheit (1°C). Action must be taken to reduce climate change. To do that, politicians must pass laws. Countries must work together.

THE ROLE OF POLITICS

Politicians and leaders create laws, such as taxes for emissions. Companies would have to pay when they release greenhouse gases. Taxes can influence which energy sources companies use.

People's politics influence which scientists and climate change studies people trust. Republicans are less likely to believe climate change is happening or feel it is a risk than Democrats are. Republican politicians tend to focus on issues such as the economy over climate change. However, the divisions

PERSPECTIVES

VOTING FOR THE CLIMATE

Scientists believe that politicians who will make laws to help stop climate change are vital to saving Earth. Riley Dunlap is a sociology professor. In a 2016 interview, he said, "I think the solution [to addressing climate change] is to elect more candidates, or more politicians, who recognize that climate change is important and action needs to be taken."

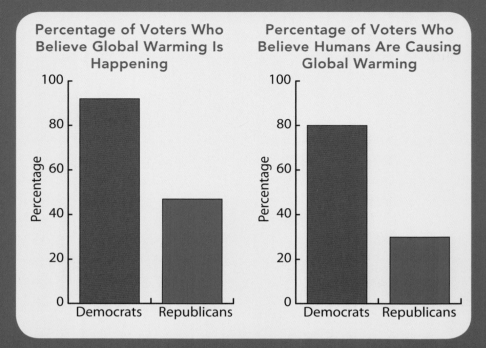

A POLITICAL DIVIDE

This graph shows the responses from an April 2019 survey. In it, 475 Democrats and 488 Republicans who were registered to vote were asked about their opinions on climate change. What do you notice about these two graphs? How do the graphs help you understand the text?

between Democrats and Republicans are not permanent. Additionally, not all Republicans are against addressing climate change. And some Democrats focus on economy before climate. But politicians must work together. They must create laws that lower carbon emissions in order to slow climate change.

STRAIGHT TO THE SOURCE

Varshini Prakash is the co-founder of the Sunrise Movement, which unites youth in the fight against climate change. She spoke at the 2019 climate strike in New York City, saying:

We've been here before as a people. In 1970, on the first Earth Day, 20 million people were in the streets. During the civil rights movement, young students and young people were arrested and took action and took risks by the tens of thousands. And that's what it takes to make change in this country.

So I need all of you to be with me here in this fight. So I want you to imagine striking not just for one day, but day after day, marching and demonstrating incessantly, even shutting down our cities and schools and businesses to stop business as usual, unless we get what we want and need as a generation.

Source: "This Is Our Time. This Is Our Future." *Democracy Now*, 23 Sept. 2019, democracynow.org. Accessed 23 Jan. 2020.

CONSIDER YOUR AUDIENCE

Adapt this passage for a different audience, such as your principal or friends. Write a blog post conveying this same information for the new audience. How does your post differ from the original text and why?

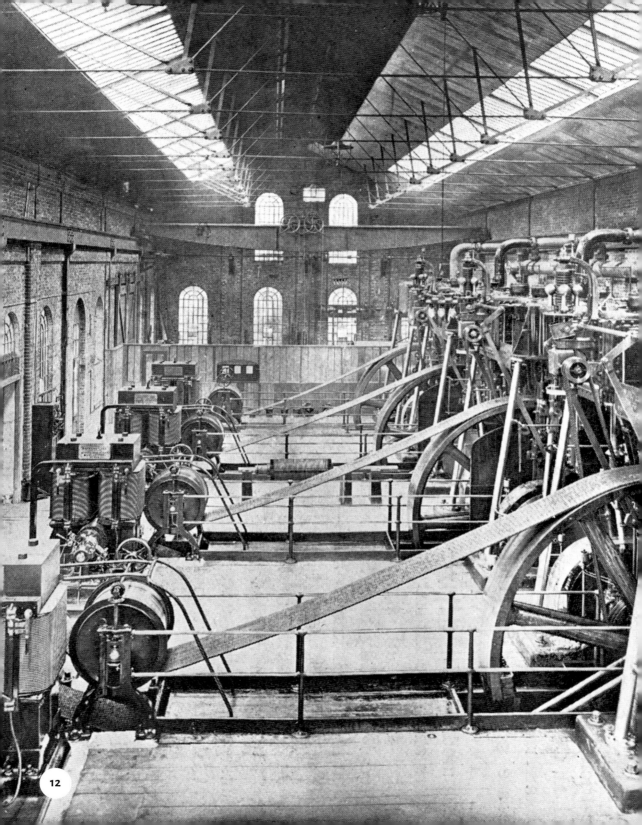

CHAPTER
TWO

RECOGNIZING A PROBLEM

Scientists have studied weather and climate for many years. Scientists have known since the 1800s how carbon dioxide affects the atmosphere. High amounts of carbon dioxide cause Earth to warm up. Still, most scientists agreed it would take thousands of years for climate change to affect people.

Then in the 1950s, some scientists noticed a warming pattern. The pattern was happening

During the Industrial Revolution, people burned coal, which generating stations turned into electricity. Burning coal releases greenhouse gases into the air.

more quickly than in the past. And it had started at the beginning of the Industrial Revolution.

People built many factories during the Industrial Revolution. Coal power plants generated energy for the factories. Those coal plants released tons of carbon dioxide into the air every year. When scientists noticed the warming trend, they began to think that the warming climate could harm people.

A WORLD CLIMATE CONFERENCE

In 1979 people gathered for the first World Climate Conference. The conference was held in Geneva, Switzerland. Farmers, businesspeople, and scientists from around the world went. They discussed Earth's climate.

The people at the conference worked in many different fields. There were 350 people from 53 countries. The people worked for 24 different organizations. At the conference, the group discussed what people knew about global warming. The group

Climate activist Luisa Neubauer spoke at the 25th Climate Change Conference held in December 2019 in Madrid, Spain.

created a document. The document explained the possible dangers of rising temperatures.

From its research, the group wrote a list of recommendations for politicians. The scientists urged countries around the world to study fossil fuel emissions. At the time, they did not know exactly

Young climate activists, including Greta Thunberg, *center*, attended an IPCC meeting in August 2019.

how emissions affected the climate. Scientists needed more information.

As a result, some countries began to form organizations. The organizations studied the climate and its changes. The studies could be used to create climate laws.

THE IPCC

After the convention in Geneva, the United Nations (UN) saw it needed to know more about climate change. The UN is an organization made up of countries around the world. The countries work together to make international agreements. The UN also creates committees to research issues. In 1988 the UN created the Intergovernmental Panel on Climate Change (IPCC).

The IPCC is not tied to any political beliefs. Its members are supposed to study climate change without bias. The panel is one of the largest climate change organizations.

PERSPECTIVES

STOP WAFFLING

In 1988 NASA scientist James Hansen spoke before the US Congress. He said that the effects of climate change were already happening. Weather patterns were changing. He urged politicians to act. Hansen said, "It is time to stop waffling so much and say that the evidence is pretty strong that the greenhouse effect is here."

The IPCC also studies how climate change affects people. Its reports help lawmakers decide how to address climate change. Lawmakers want to make laws that help people. Some want to limit the harmful effects of climate change.

In 1990 the IPCC released its first report. It called for a global treaty among countries. The countries would agree to limit the effects of climate change. The report noted that fossil fuels add greenhouse gases to the air. It recommended that countries make stricter emissions laws.

RESISTANCE

Oil and gas companies such as ExxonMobil did not want laws to limit emissions. So ExxonMobil created a series of advertisement and lobbying campaigns. The campaigns made the public doubt the accuracy of climate science.

Both the public and politicians began to doubt climate science. Politicians wondered whether climate

Fossil fuel companies would be negatively affected by emission laws.

change was an important issue. Changes to laws could cost billions of dollars and might mean raising taxes. Politicians know most people do not want higher taxes.

Researchers believe the campaigns from fossil fuel companies likely slowed changes to policies concerning climate change. ExxonMobil said it only wanted to make sure the studies were right. Company leaders argued it did not make sense to spend so much money if there was no reason to be concerned. But scientists insisted that climate change was an issue.

THE KYOTO PROTOCOL

In 1997, 41 countries and the European Union met in Kyoto, Japan. They signed an agreement to limit greenhouse gas emissions. The agreement was called the Kyoto Protocol. Each country had its own goals to reduce emissions.

THE GOVERNMENT'S RESPONSIBILITY

In 2007 the US Supreme Court passed a ruling about carbon emissions. It stated the Environmental Protection Agency (EPA) could limit greenhouse gas emissions. The EPA sets guidelines to protect the environment. The ruling said carbon dioxide endangers the public. Therefore, the EPA could and should regulate carbon dioxide emissions.

A representative for the United States signed the Kyoto Protocol. But few Democratic or Republican congress members supported it. Congress never approved the protocol. The protocol was set to go into effect in 2005 in other countries. That year President George W. Bush pulled out of the agreement. Bush said the protocol would hurt US families and the economy.

FURTHER EVIDENCE

Chapter Two talks about how scientists raised awareness about climate change. Identify one of the chapter's main points. What evidence does the author provide to support this point? The website below discusses how people are learning to live with climate change. Find a quote on this website that supports the main point you identified. Does the quote support an existing piece of evidence in the chapter? Or does it offer a new piece of evidence?

LIVING WITH CLIMATE CHANGE

abdocorelibrary.com/climate-change-politics

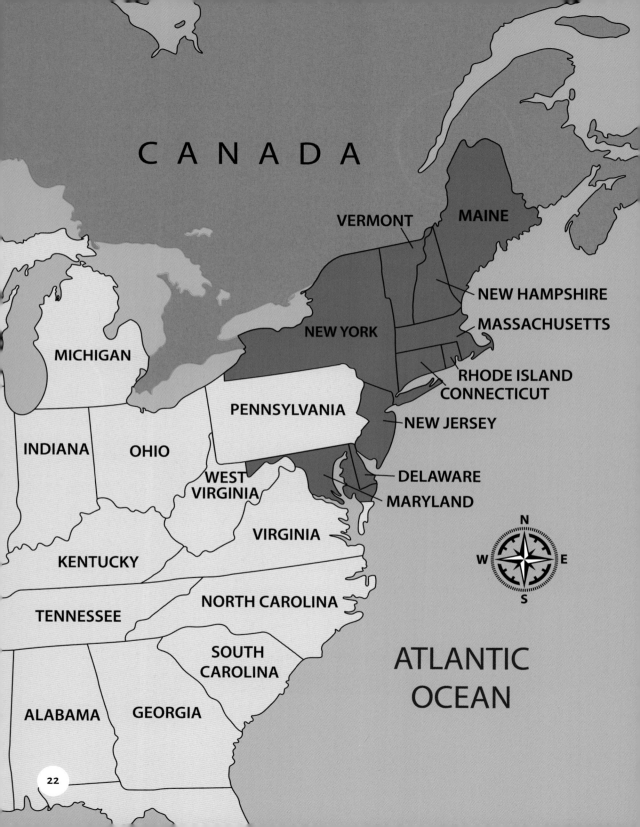

CHAPTER THREE

PIECE BY PIECE

In the early 2000s, more people were beginning to believe humans were causing climate change. More people also believed climate change was threatening the planet. Scientific studies confirmed this. Many states made positive changes. For example, in 2005 several states on the East Coast created the Regional Greenhouse Gas Initiative (RGGI). The states worked together to reduce emissions. There was bipartisan work to reduce emissions all across the country.

By 2020, 10 states had joined together under the RGGI to lower their greenhouse gas emissions.

New York's former governor George Pataki, left, gave a joint speech on the RGGI with former California governor Arnold Schwarzenegger in 2006. Pataki started the discussions that led to the RGGI.

However, in 2009 and 2010, the US political response to climate change became more divided. Fewer Republican and Democratic politicians agreed on what to do to reduce climate change.

EXXONMOBIL

Between 1999 and 2005, Lee Raymond was the chief executive officer (CEO) of ExxonMobil. During this period, the company funded many political organizations. Some of those organizations work to deny the role humans play in climate change.

But in 2006, Raymond left ExxonMobil. Rex Tillerson became the new CEO. Tillerson had a different view on climate change. Tillerson agreed with scientists. He said climate change was caused by greenhouse gases. He agreed that people were making climate change worse. Tillerson said action needed to be taken.

Many environmental activists thought this would be an important turning point. ExxonMobil is a powerful fossil fuel company. Activists hoped other companies would follow it and limit emissions.

However, changes in ExxonMobil's policies have been slow. In 2012 Tillerson said he did not want to act too quickly. He suggested other companies should also take it slow. Tillerson said there were enough oil and gas reserves to last a long time. He added that the planet adapted to climate change in the past.

Al Gore has continued to speak out about climate change. He believes it is the most serious challenge the world has ever faced.

He believed it would adapt again. Scientists argue that living things will not be able to adapt. The changes are happening too quickly.

CLIMATE CHANGE IN FILM

During this same period, public opinion continued to change. In 2006 a documentary about climate change was released. Former vice president Al Gore wrote and starred in the film. It was called *An Inconvenient Truth*.

In the film, Gore outlined the dangers of climate change. The film was a success. But it also received criticism. Much of the criticism came from climate change skeptics. One film responded to the documentary. The short film was called "Energy." It countered Gore's film by saying carbon was a source of life. It said carbon was not a pollutant.

"Energy" was not wrong about carbon. Carbon is a source of energy. Without it, life could not exist. However, in large quantities, carbon and carbon dioxide create harmful effects.

Some people tried to keep Gore's film out of schools. One teacher in the United Kingdom took Gore and his film to court.

GORE'S ENERGY USES

After *An Inconvenient Truth*, Al Gore was criticized for his choices. Some people said he flew in a private jet. Air travel is responsible for a lot of carbon emissions. Gore said he does not fly in a private jet. But people also noticed Gore's home. Gore lives in a large house. It takes a lot of energy to heat and power it. People wondered if Gore really believed the things he said in the film. Gore later made his home more energy efficient. He also started using some renewable energy, such as solar energy.

The teacher claimed the film exaggerated the science of climate change.

The court ruled the science in the film was mostly accurate. As a result, it should be allowed to be taught in schools. However, the judge agreed that some parts of the film were exaggerated, so notes for teachers should be provided. Teachers around the world used the film. It helped spark discussions about what can be done to slow climate change.

THE TURNING POINT?

By 2008 most politicians agreed that climate change was caused by human actions. There was support across parties for limiting carbon emissions. Both Democratic and Republican politicians supported stricter greenhouse gas laws.

Lawmakers began writing a bill called the American Clean Energy and Security Act of 2009. The bill was created to reduce air pollution. It also would have made the United States use more renewable energy. The bill

Public opinion on climate change has varied over the years. However, more Americans are recognizing it as an important issue and believe the government should be doing more to combat climate change.

passed the House of Representatives, but it never managed to pass the Senate. There was a Democratic majority at the time. Even so, the Senate leader said not enough people were willing to pass it. Lawmakers did not write another bill.

PERSPECTIVES
NOT A JOKE
President Barack Obama ran for re-election in 2012. During the campaign, his Republican opponent, Mitt Romney, joked about climate change. At the Democratic National Convention in August, Obama responded. He said, "Climate change is not a hoax. More droughts and floods and wildfires are not a joke. They are a threat to our children's future."

In 2009 public opinion shifted. The Pew Research Center held two surveys. One was in 2008, the other in 2009. Results from the surveys showed fewer people believed climate change was caused by human actions. In 2008, 47 percent of people believed climate change was caused by human actions. But in 2009, that number dropped to 36 percent. In the same survey, public belief that climate change existed dropped from 71 percent to 57 percent.

Many events influenced the change in public opinion. The Great Recession (2007–2009) caused many people to lose their jobs and homes. People were not focused on climate change. Additionally, a 2009 report worked to discredit climate science. Two climate change denial organizations created the report. The report stated that a warming climate would benefit people. It wouldn't cause harm.

In 2010 Americans voted in midterm elections. The elections brought a Republican majority to the House of Representatives. Many of the new lawmakers did not want to make laws to limit climate change. This led to another delay in climate laws.

EXPLORE ONLINE

Chapter Three talks about how public opinion about climate change has shifted over time. The article at the website below goes into more depth on this topic. Does the article answer any of the questions you had about people's views?

CLIMATE BASICS FOR KIDS

abdocorelibrary.com/climate-change-politics

CHAPTER FOUR

AN UNCERTAIN FUTURE

The divide between Democratic and Republican views grew. It became even wider after President Barack Obama's re-election in 2012. Tensions about how to fix climate change grew even more during the 2016 election. The United States elected Donald Trump as president. Trump denied climate change was an issue.

The divide was strongest among middle-aged and older adults. Young people tended to be in closer agreement about

Barack Obama met with CEOs from US companies to discuss how to reduce climate change.

climate change. In part because of young people's influence, Republicans began to discuss climate change and how to fight it.

By the late 2010s, politicians were using scientific studies to create laws. For example, in 2018 Canada used scientific studies to create a tax. This tax was for carbon-emitting products such as gasoline. Studies have shown the effectiveness of carbon taxes. For example, Sweden's

PERSPECTIVES

LEAVING THE PARIS AGREEMENT

The Paris Agreement is an international agreement made in 2015. Its goal is to lower carbon emissions. All but three countries had signed the agreement by 2017. In that year, President Donald Trump pulled out. He believed the agreement cost the United States too much money. However, a future US president could reverse that decision. French president Emmanuel Macron was disappointed with Trump's decision. He said, "I tell you firmly tonight: we will not renegotiate a less ambitious [agreement]. There is no way. . . . Don't be mistaken on climate: there is no plan B because there is no planet B."

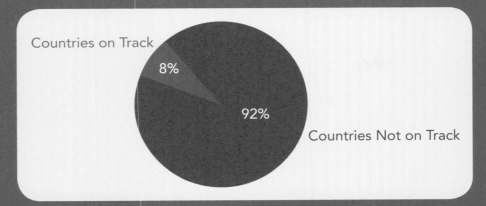

PARIS AGREEMENT LAWS

By 2018, 197 countries had signed the Paris Agreement. When the countries signed the agreement, they promised to limit greenhouse gas emissions. However, only 16 countries were on track to meet their goals. This chart shows the percentage of countries that were and were not on track with their goals. What does this tell you about the view many politicians in other countries have about climate change?

emissions fell by 23 percent between 1990 and 2015. Politicians can look at other states and countries. They can use other laws as models.

However, politicians also rely on donations from lobbyists. Lobbyists are people working for a company. They encourage politicians to act in the company's best interests. Lobbyists give a lot of money to politicians. In return, the politicians pass laws that help a company.

In 2019 climate change activist Greta Thunberg was nominated for a Nobel Peace Prize for her work, though she did not win.

They block laws that would hurt a company. Powerful lobbyists work for coal and gas companies.

GRETA THUNBERG

Young people are spurring political change. Greta Thunberg is a young Swedish climate activist. In 2018 Greta sat in the Swedish Parliament every day for three weeks. She missed school to protest the lack of action to fight climate change. Greta also founded the movement Fridays for Future. She suggested students protest on Fridays. Fridays for Future is also a way to share about upcoming climate protests around the world.

In 2019, at age 16, Greta met with US politicians. She spoke in front of the US Senate. She urged lawmakers to act against climate change. Greta Thunberg was named *Time* magazine's Person of the Year in 2019. She inspired many young people to take action. Her work helped create the Global Climate Strike. People felt empowered to talk to politicians. Young people marched in protests.

BEING HEARD

In December 2018, Alexandria Villaseñor began spending her Fridays in front of the UN building in New York City. She was protesting the lack of response to climate change. Alexandria turned 14 years old in 2019. Most Fridays she was alone. But in September 2019, Greta Thunberg and hundreds of other students joined Alexandria. They marched down the street together. As a result, the president of the UN General Assembly asked to speak with Alexandria.

POWER TO INDIGENOUS PEOPLES

Indigenous peoples around the world work to protect lands. Much of the lands on which indigenous peoples live or used to live are forested. Some indigenous peoples help keep forests healthy. For example, members of the Coquille Tribe in Oregon work to keep plant life and wildlife diverse in their forests. Caring for forests is important. Forests are home to many different species of plants and animals. Forests also act like a sponge for carbon dioxide. Trees and other plants take in carbon dioxide. They give off oxygen. This process lowers the amount of carbon dioxide in the air.

Indigenous peoples have had little recognition for the important role they play in slowing climate change. That began to change in 2019. The IPCC released a report. The report acknowledged the work indigenous peoples do to help the world stay healthy. During the 1800s and 1900s, many indigenous peoples in the United States were forced off their lands. The IPCC report recommended rights be given back to

Some states have invested in alternative sources of energy, such as wind energy, which do not produce greenhouse gases that harm the environment.

indigenous communities. That included land rights. The IPCC said this would help keep the land healthier. Politicians have the power to give land rights back to indigenous peoples.

THE FUTURE OF CLIMATE CHANGE AND POLITICS

In order to reduce climate change, laws must change. In the United States, more Democratic and Republican politicians are focusing on climate action.

In 2019 US lawmakers began working on a Green New Deal. The Green New Deal would be a series of policies to slow climate change. One solution was moving entirely to renewable energy by 2030. As of

2020, legislation for the Green New Deal had not been presented to Congress. Climate laws still face resistance. Many people did not like the Green New Deal because it was expensive.

Some of the biggest growth in renewable energy has been in traditionally Republican-leaning states. For example, a study showed Kansas tripled its wind power in just four years. Additionally, North Dakota produced the most wind energy per person living in the state. Researchers said the study showed climate action does not have to be Republican versus Democrat. People from both parties care about jobs and clean air. Renewable energy provides construction jobs in rural areas.

There is still a long way to go before politicians are united about climate change. But if people are going to slow climate change, politicians must work with one another. They must learn from scientists. They must work toward a cleaner future.

STRAIGHT TO THE
SOURCE

Paula Garcia is a climate scientist. In 2018 she gave an interview about how political decisions affect renewable energy research.

> At the beginning of this year [2019], the administration of President Trump put in place solar tariffs, and the uncertainty of what was going to happen with that decision created a big problem for the solar industry. . . .
>
> We have lost more than 10,000 jobs in the solar industry because of that decision. . . . Then the other challenge is that I see that this administration wants to help the fossil fuel industry with an emphasis on coal. . . . So, it's very strange for me to see that this administration is trying to do that because it is going to have a bad impact for the health of the people that live in this country.
>
> Source: MacDonald, Colleen. "Getting Excited about Energy." *Union of Concerned Scientists*, 21 Aug. 2018, ucsusa.org. Accessed 6 Dec. 2019.

BACK IT UP

The author of this passage is using evidence to support a point. Write a paragraph describing the point the author is making. Then write down two or three pieces of evidence the author uses to make the point.

FAST FACTS

- People use protests such as the 2019 Global Climate Strike to try to convince politicians to make laws to slow climate change.

- The laws that politicians create are important to slowing climate change.

- The World Climate Conference was held in 1979. Attendees discussed global warming and its possible dangers.

- The United Nations created the Intergovernmental Panel on Climate Change (IPCC) in 1988. The IPCC studies climate change and how it affects people.

- Former vice president Al Gore's documentary *An Inconvenient Truth* came out in 2006. This film inspired many to take climate change seriously.

- The American Clean Energy and Security Act of 2009 failed to make it through the US Senate.

- Carbon taxes are becoming a common way in other countries to slow climate change.

- Greta Thunberg is a young climate activist from Sweden. She speaks with politicians around the world to try to persuade them to take climate change seriously.

- Many indigenous peoples care about keeping their lands healthy. The IPCC stated in 2019 that their work helps keep the world healthy in the face of a warming climate.

STOP AND THINK

Tell the Tale

Chapter One of this book discusses the Global Climate Strike in 2019. Imagine you are in the crowd during the strike. Write 200 words about what you see around you. Why might a climate strike make politicians listen?

Surprise Me

Chapter Two discusses scientists' first call to politicians to act on climate change. After reading this book, what two or three facts about climate change and politics did you find most surprising? Write a few sentences about each fact. Why did you find each fact surprising?

Dig Deeper

After reading this book, what questions do you still have about climate change laws? With an adult's help, find a few reliable sources that can help you answer your questions. Write a paragraph about what you learned.

You Are There

Chapter Three discusses how political leaders have gone back and forth on climate change policies. Imagine you are a scientist in a meeting with a political leader. Write a letter to the politician describing why climate action is important to you. Be sure to add plenty of details.

GLOSSARY

activist
a person working to change a law

bipartisan
involving lawmakers of different political parties

campaign
work done to achieve a goal, often in political elections

economy
the money and resources a country has

fossil fuel
fuel created over millions of years from decaying plant and animal matter

legislation
a proposed law written by lawmakers

lobbyist
a person who works to convince a legislator to act in a particular way that benefits the person or a corporation

tariff
a fee that a government charges on goods brought into the country for sale

treaty
an agreement between or among nations

ONLINE RESOURCES

To learn more about climate change and politics, visit our free resource websites below.

Visit **abdocorelibrary.com** or scan this QR code for free Common Core resources for teachers and students, including vetted activities, multimedia, and booklinks, for deeper subject comprehension.

Visit **abdobooklinks.com** or scan this QR code for free additional online weblinks for further learning. These links are routinely monitored and updated to provide the most current information available.

LEARN MORE

Harris, Duchess. *Environmental Protests*. Abdo Publishing, 2018.

London, Martha. *The Effects of Climate Change*. Abdo Publishing, 2021.

INDEX

American Clean Energy and Security Act of 2009, 28–29

Bush, George W., 21

Democrats, 6, 9, 10, 21, 24, 28–29, 30, 33, 39, 40

ExxonMobil, 18–20, 24–25

Garcia, Paula, 41
Global Climate Strike, 5–6, 11, 37
Gore, Al, 26–27
Green New Deal, 39–40

Inconvenient Truth, An, 26–28
indigenous peoples, 38–39
Industrial Revolution, 7, 14
Intergovernmental Panel on Climate Change (IPCC), 17–18, 38–39

Kyoto Protocol, 20–21

Macron, Emmanuel, 34

New York City, 6, 11, 37

Obama, Barack, 30, 33

Paris Agreement, 34, 35

Regional Greenhouse Gas Initiative (RGGI), 23
Republicans, 9–10, 21, 24, 28, 30, 31, 33–34, 39–40

Thunberg, Greta, 6, 36–37
Tillerson, Rex, 25
Trump, Donald, 33, 34, 41

United Nations (UN), 17, 37

World Climate Conference, 14–16

About the Author

Martha London writes books for young readers. When she isn't writing, you can find her hiking in the woods.

St Croix Falls Library
230 S Washington St
St Croix Falls WI 54024